SELFIES FROM MARS

The True Story of Mars Rover Opportunity

Fact File

This little rover is named Oppy, in honor of Opportunity's nickname. Look for Oppy's Fact Files throughout the book for bonus information!

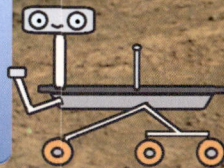

Katie L. Carroll

Dedication: To all the explorers out there, big and little, human and robot.

Photo Credits

Lyrics Rights

The information in this book is true and complete to the best of our knowledge. All recommendations are made without guarantee on the part of the author or publisher. The author and publisher disclaim any liability in connection with the use of this information.

Library of Congress Cataloguing-in-Publication Data
Name: Carroll, Katie L., author.
Title: Selfies from Mars: the true story of Mars rover Opportunity / Katie L. Carroll.
Identifiers: LCCN 2022923896 (print)   ISBN 9781958575017 (print)   ISBN 9781958575000 (ebook)
Subjects: 1. Mars (Planet)—Exploration—Juvenile literature. 2. Roving vehicles (Astronautics)—Juvenile literature.
LC record available at https://lccn.loc.gov/2022923896

Published by Shimmer Publications, LLC
Milford, Connecticut
Visit the author's website at www.katielcarroll.com

On a summer night in 2003, a Delta II rocket launched from Cape Canaveral Air Force Station, Florida, in a blast of vapor.

Folded up inside like an origami crane, the Mars Exploration Rover (MER) Opportunity was on its way to the Red Planet.

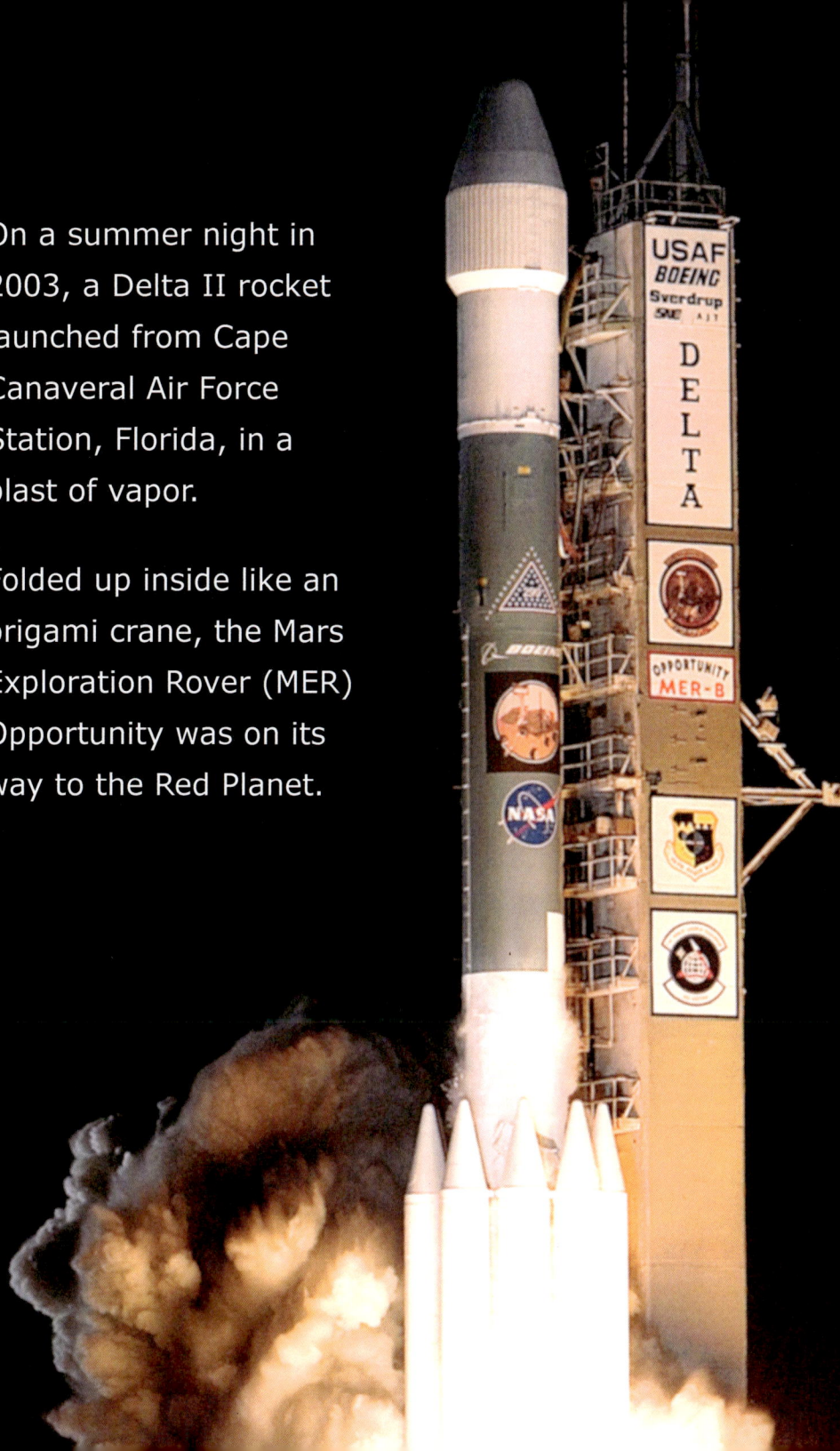

Cheers from NASA scientists and engineers rose up from Earth. It was a moment of relief as several attempts at launch had been scrubbed due to bad weather or faulty equipment.

Three weeks earlier, Spirit—Opportunity's twin—had taken flight.

A fully assembled Opportunity spends time in the Spacecraft Assembly Facility at NASA's Jet Propulsion Laboratory in California shortly before departing for the Kennedy Space Center in Florida.

Launch: Complete

The MER team at the Jet Propulsion Laboratory (JPL) in Pasadena, California, had already put in years of hard work to design and build these twin rovers.

Yet the launches were only the beginning of the rovers' journeys.

Top: Workers check equipment, while Opportunity sits atop the deployed lander, its solar arrays and wheels tucked in tight. Bottom left: The parachute is tested in a wind tunnel at NASA's Ames Research Center in California. Bottom right: Members of the MER team pose with the rover.

Interplanetary travel is always a risk. In the late 1990s, several previous missions to Mars had failed.

Now Opportunity and Spirit were headed for opposite ends of an alien world, never to meet again.

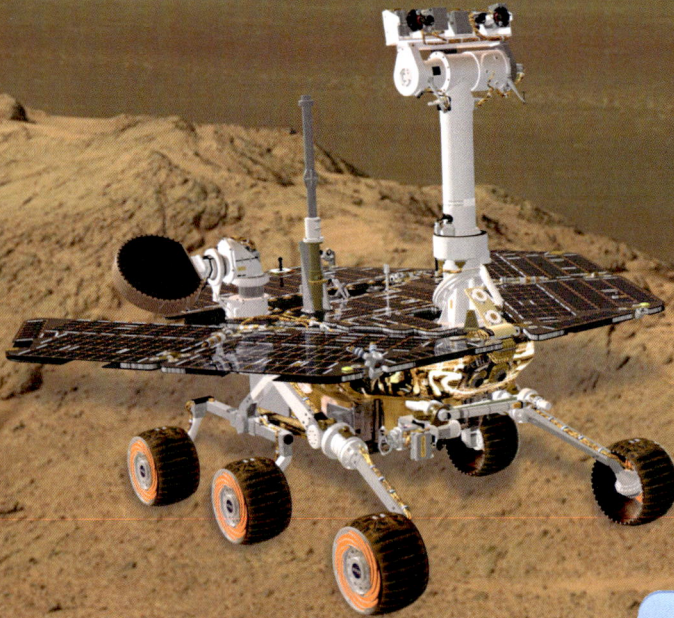

Fact File

Opportunity's landing site of Meridiani Planum and Spirit's of Gusev Crater are both near the equator of Mars, but they are halfway across the planet from each other.

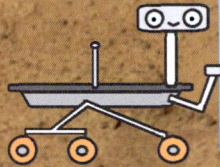

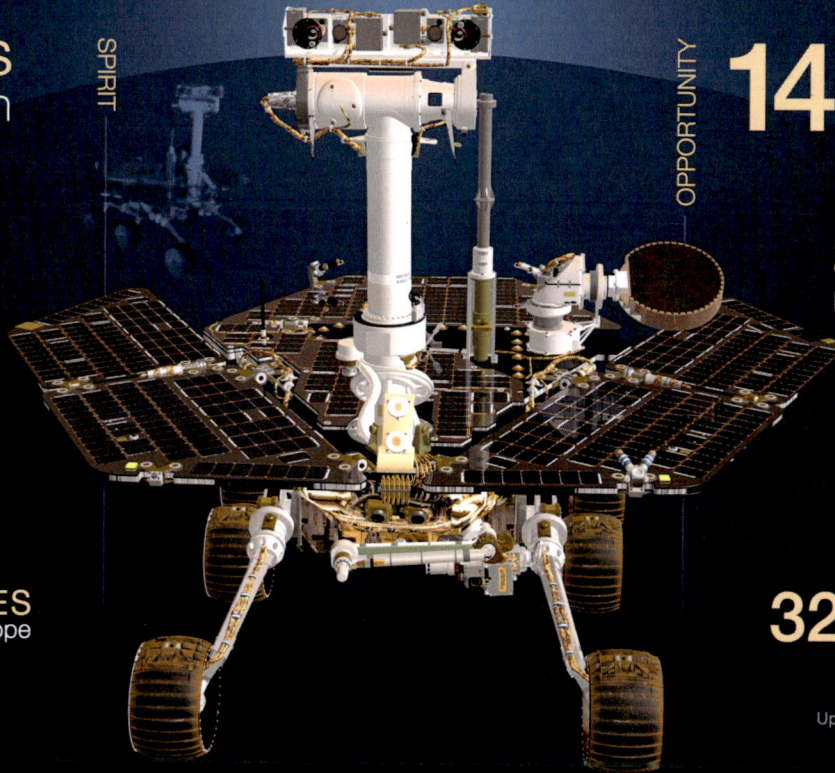

Spirit and Opportunity
BY THE NUMBERS

SPIRIT

OPPORTUNITY

6 YEARS
lifespan

14+ YEARS
lifespan

124,838
raw images

217,594
raw images

4.8 MILES
traveled

28 MILES
traveled

30 DEGREES
steepest slope

32 DEGREES
steepest slope

Updated February 4, 2019

When Spirit and Opportunity launched, no one knew how long the rovers would function. The proof of their resilience is summarized in the mission numbers above.

Mars is a cold, dry place—unsuitable for human life.

It is also a challenging world for robot rovers. Giant dust storms can cover the entire planet and last for months. These storms block out the sun, the life source for solar-powered rovers.

Fact File

The thin Mars atmosphere contains about 95% carbon dioxide with very little oxygen. At -80° Fahrenheit (-62° Celsius), the average temperature is colder than Earth's South Pole.

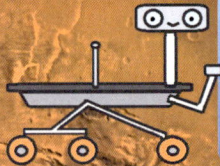

Even when the sun is out, Mars is a strange world.

The distinct rust color of the iron-rich surface gives the Red Planet its nickname. Tiny dust particles high in the atmosphere paint the sunsets blue.

Using its panoramic camera, MER rover Spirit snapped this view of the blue Martian sunset.

Conditions on Mars weren't always so strange and hostile. Life may have existed on ancient Mars. Billions of years ago, its atmosphere was much thicker and warmer.

Scientists believe water once flowed in great valleys on the surface. These conditions could have supported tiny life forms.

Today, most of the water exists as ice, much of it in the planet's polar regions.

An artist's rendering of ancient Mars with a thick, Earthlike atmosphere and liquid water.

Opportunity's mission was to find evidence of water on the now dry planet.

For seven months, Opportunity traveled in the vast space between Earth and Mars. Then in January 2004, the rover made its final approach.

An artist's rendering of modern Mars with a thin atmosphere and cold, dry conditions.

Opportunity plummeted toward Mars at dizzying speeds, hitting the atmosphere at 12,000 mph (5,364 m/s). An outer heat shield protected the inside of the lander from temperatures as hot as the surface of the sun.

At 1,000 mph (447 m/s), the parachute deployed and rapidly slowed the craft. Then the heat shield dropped off and the lander was suspended on a rope from the back shell.

Seconds before crash landing on Mars, the airbags inflated, retro rockets on the back shell further slowed the craft, and the rope released the lander.

On the surface of Mars at last, Opportunity bounced and bounced and bounced. With each bounce, the airbags left behind distinct marks on the planet's surface.

The rover, safely tucked inside the airbags, rolled and rolled and rolled. It came to a stop in a small crater: A hole in one!

With Spirit having landed a few weeks earlier, the MER team once again celebrated.

Entry, Descent, & Landing: Complete

Fact File

Entry, Descent, and Landing (EDL) takes six minutes, but communications take ten minutes to reach Earth. There is nothing to do but wait, so scientists refer to EDL as "six minutes of terror."

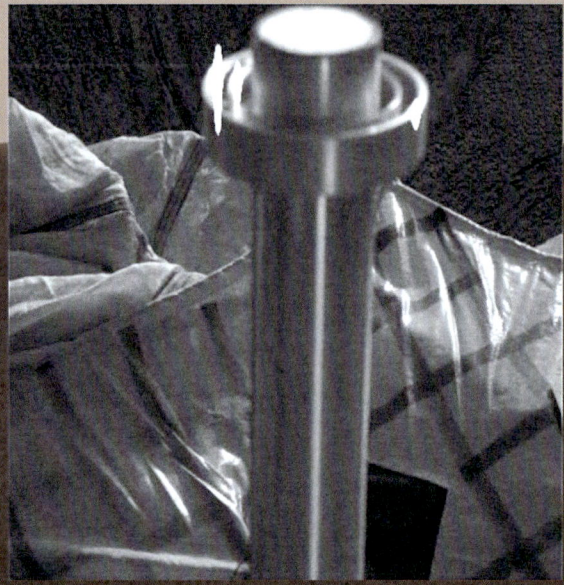

A series of images taken by Opportunity's panoramic camera on solar day (sol) 1.

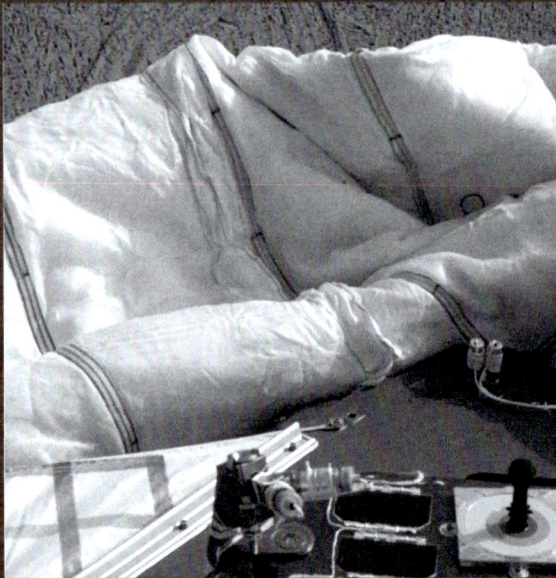

Opportunity was not yet safe. Many things had to happen on those first Martian days, called sols.

If the solar arrays didn't deploy, the rover would never charge. If the camera mast didn't rise, it wouldn't take pictures. If the antenna didn't unfold, it wouldn't send or receive data. If the body didn't jack up and the wheels didn't pop out, it would never move.

Fact File

A day on Mars is called a sol. It's the time it takes the planet to rotate once on its axis. At 24 hours, 39 minutes, and 35 seconds, a Martian sol is slightly longer than an Earth day.

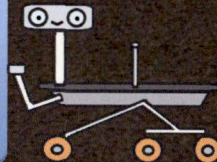

The MER team munched on peanuts—a good-luck tradition at JPL—while anxiously waiting to hear from the rover over NASA's Deep Space Network (DSN). Data came in over the next sols, indicating that Opportunity's sensitive equipment had weathered the landing.

On sol 7 (January 31, 2004 on Earth), the team played the song, "Born to Run" by Bruce Springsteen, to rouse the rover. Then Opportunity's six wheels rolled off the fabric ramps and hit Martian soil for the first time.

The rover was off to explore!

Deployment & Egress: Complete

Opportunity's landing site, dubbed Eagle Crater, with the rover's wheel prints showing where it drove around the empty lander.

This Goldstone 111.5-foot antenna at a facility in California's Mojave Desert is part of the DSN. Along with facilities in Australia and Spain, the DSN keeps in constant contact with spacecraft as Earth rotates.

Opportunity and Spirit were designed to be golf-cart-sized geologists.

Their panoramic cameras were "eyes." Their computers were "brains" with bravery settings to help them decide whether to drive over objects or around them.

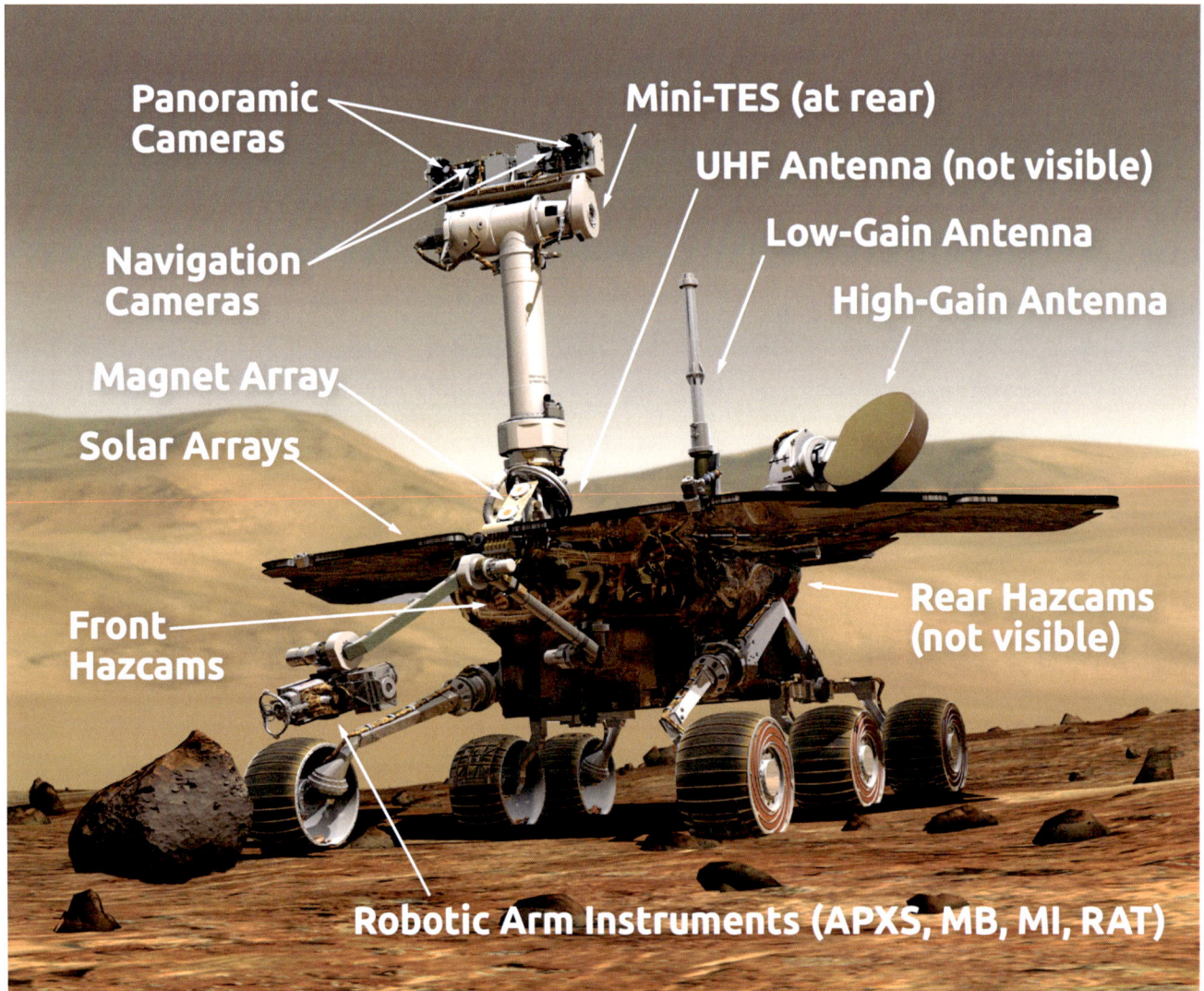

Panoramic Cameras

Mini-TES (at rear)

UHF Antenna (not visible)

Navigation Cameras

Low-Gain Antenna

High-Gain Antenna

Magnet Array

Solar Arrays

Rear Hazcams (not visible)

Front Hazcams

Robotic Arm Instruments (APXS, MB, MI, RAT)

The rovers each had a robotic arm with three joints similar to a shoulder, elbow, and wrist.

At the end of the arm were instruments that included the first interplanetary power tool. Called the RAT—short for rock abrasion tool—it scraped alien rock to examine the composition.

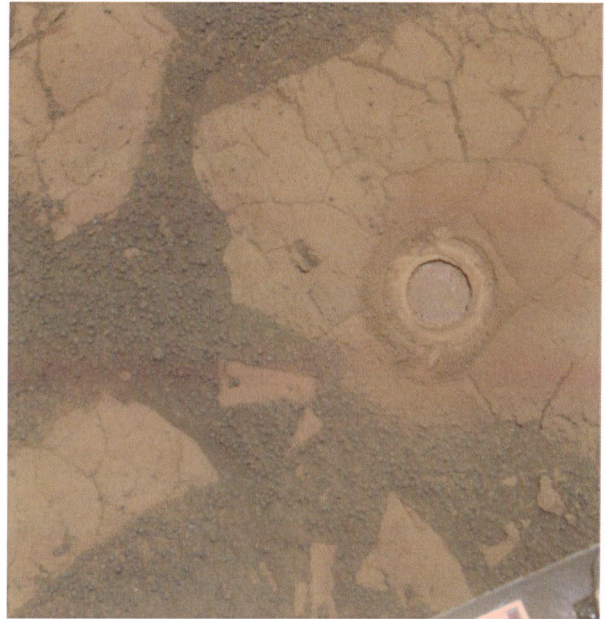

Top: Opportunity's RAT creates a circular mark to study the layered bedrock. Bottom: Instruments on the robotic arm are poised to take measurements of a trench dug in the sand.

Immediately from sol 1, Opportunity was sending images of its surroundings.

The MER team could hardly believe what they were seeing!

The pictures showed a layered outcrop of bedrock, a type of rock never closely observed on Mars. Each layer was like going back in time to discover the planet's history.

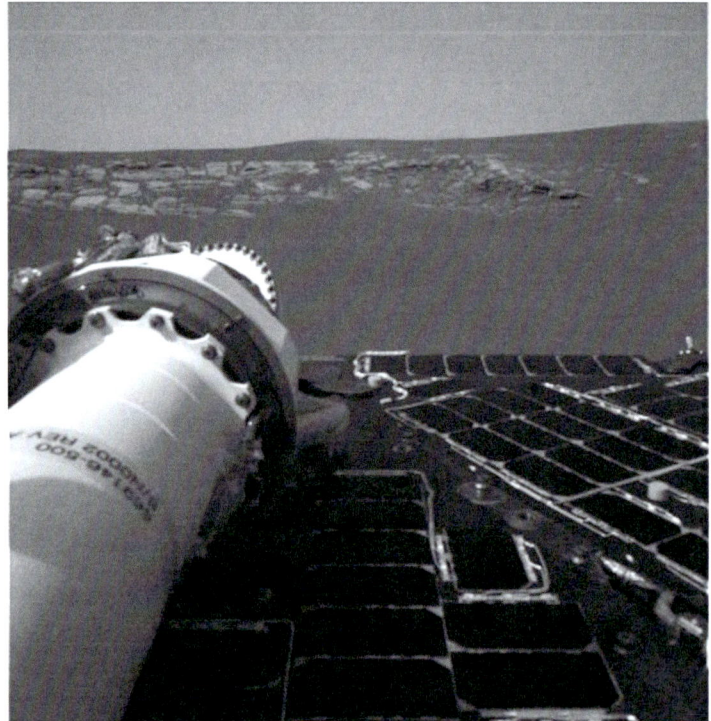

Top: Opportunity's solar panels and mast with a background of a bright outcrop of layered rock.
Bottom: A close-up of layered rock found in Eagle Crater where Opportunity landed.

There were also curious tiny spheres on the ground all around the rover. The team used the nickname "blueberries" for the little gray balls, which they suspected were hematite. Hematite is a mineral that forms in liquid water and contains iron oxide.

On sol 46 (March 11, 2004 on Earth), Opportunity gathered data on a cluster of blueberries and confirmed the team's suspicions that the balls were hematite. With that and expert analysis of the layered rock, the team concluded that this area of Mars once contained liquid water.

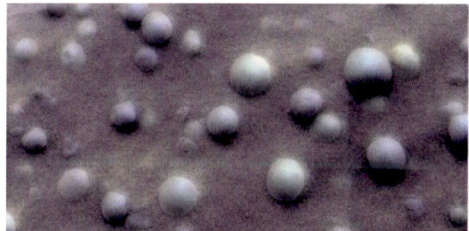

Evidence of Water: Complete

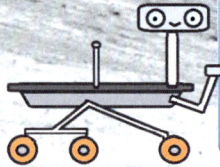

Fact File

Small dust devils, like the one pictured above captured by Opportunity's navigation camera, helped to extend the rover's mission by blowing dust off the solar panels.

These findings alone would have been enough to call the mission a scientific win. Had Opportunity merely lasted the 90 sols it was designed for, the mission would have been considered a rousing success.

Yet Opportunity traveled on for those 90 sols and many more. For years, the rover revealed the strange beauty and harsh conditions of Mars, helping humans experience a faraway planet as never before.

It took pictures that showed the vast Mars landscape; it left wheel tracks in the red Martian soil; it climbed mountains, explored craters, and endured storms.

Opportunity also took a rover selfie!

Although the rover had captured images of itself using the panoramic camera, Opportunity had never attempted a self-portrait of this kind.

This self-portrait is a series of images taken by Opportunity's panoramic cameras, which were mounted on the mast. The images were stitched together for this view of the rover from above.

Between sols 5000 (February 26, 2018 on Earth) and 5006, Opportunity used its microscopic imager, mounted on the end of its arm, to take a series of pictures to create a true selfie. The unique image gave Earthlings a whole new view of the rover.

The microscopic imager that took the shots for the selfie was designed for examining rocks and soil up close, which is why Opportunity appears out of focus in this image.

Six years after landing, Spirit was still on the opposite side of the planet from its twin, Opportunity. On sol 2160 of the mission, Spirit was stuck in soft soil with its wheels beginning to fail. The rover settled in for the Martian winter.

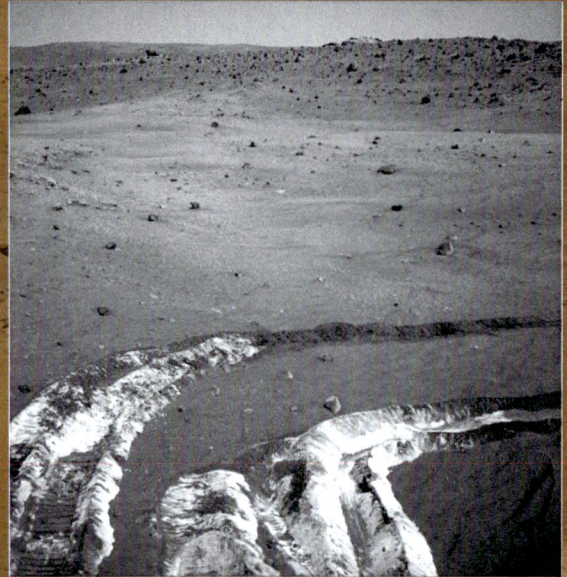

Left: Spirit's microscopic imager captured tiny pits on a rock that contains hematite. Right: Spirit's wheel marks uncovered brightly colored soil where the rover's spectrometer detected salts.

Spirit attempted to tilt into a position to optimize sun exposure so the batteries would continue to charge. But it wouldn't be enough to sustain the rover.

Spirit's final communication with Earth was on sol 2210 (March 22, 2010 on Earth).

For more than a year, the team at JPL attempted to rouse the rover before calling it the end for Spirit.

Spirit's Mission: Complete

Back on the other side of Mars, Opportunity kept on going. For more than 14 years, it roved along the Martian surface, setting driving records, snapping more than 200,000 images, and making ground-breaking discoveries about the Red Planet.

The scientific explorations of Opportunity and Spirit taught the MER team about the current conditions of Mars as well as its history. Their resilience proved what it takes for roving robots to survive on Mars and paved the way for future rovers like Curiosity and Perseverance.

Top: Opportunity utilizes the position of the sun to capture its own shadow. Bottom: Opportunity looks back at its wheel tracks after being stuck in the sandy soil for more than five weeks.

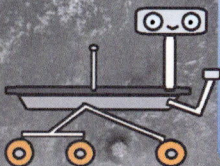

Endurance crater

Victoria crater

Santa Maria crater

Perseverance Valley

Endeavour crater

5 km

Fact File

Clocking in at 28 miles (45 kilometers), Opportunity posted a record as the vehicle that traveled the farthest on another world. This record endures three years after the mission's end, even as bigger, faster rovers explore Mars.

On sol 5111 (June 10, 2018 on Earth), Opportunity sent data to scientists that they poetically interpreted to mean "my battery is low and it's getting dark."

A giant dust storm was approaching, soon to cover the whole planet.

Then Opportunity went silent.

The dust storm had already blocked the sun when Opportunity sent this partial image. The black at the bottom shows where the image transmission was cut off.

The Mars Orbiter Camera on the Mars Global Surveyor took these images of the planet in 2001. A relatively clear Mars is shown on the left image, while on the right, the planet is almost entirely covered by a dust storm.

Fact File

Dust storms on Mars can reach global scale within a few days' time. Storms that block the sun for long periods of time and cause dust to settle on a rover's solar panels make it difficult for the rover to power back up.

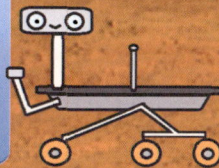

But hope remained. Opportunity had shown its capacity for survival before. Back on sol 1233 (July 13, 2007 on Earth), powerful storms raged along Mars's southern hemisphere, covering Opportunity's solar panels with dust.

The rover waited it out by reducing activity to save power, which had dipped to its lowest levels of the mission. The dust settled and blew away, the skies cleared, and Opportunity powered back up.

When the dust cleared on this most recent storm, Opportunity did not respond. One month passed, then two…then six. The team at JPL sent more than a thousand recovery commands to the rover.

Opportunity remained silent.

As the 2018 storm begins to clear, Opportunity is a barely visible dot in the area marked by the small white square. This image of Perseverance Valley was taken by the Mars Reconnaissance Orbiter.

Finally, on sol 5353 (February 12, 2019 on Earth), more than seven months after last hearing from the rover, the team at JPL said goodbye to Opportunity.

They ate peanuts and sent a farewell message. It was a love song, a recording of Billie Holiday singing the 1944 hit, "I'll Be Seeing You." The final lyrics read:

I'll find you in the morning sun
And when the night is new
I'll be looking at the moon
But I'll be seeing you

The final call came in, "MER project off the net."

The team at JPL once again cheered for the robot rover they had affectionately called Oppy. Then the room fell silent.

John Callas, MER project manager, calls into the DSN for the last time of Opportunity's mission. The call was made from mission control at NASA's Jet Propulsion Laboratory.

Opportunity's Mission: Complete

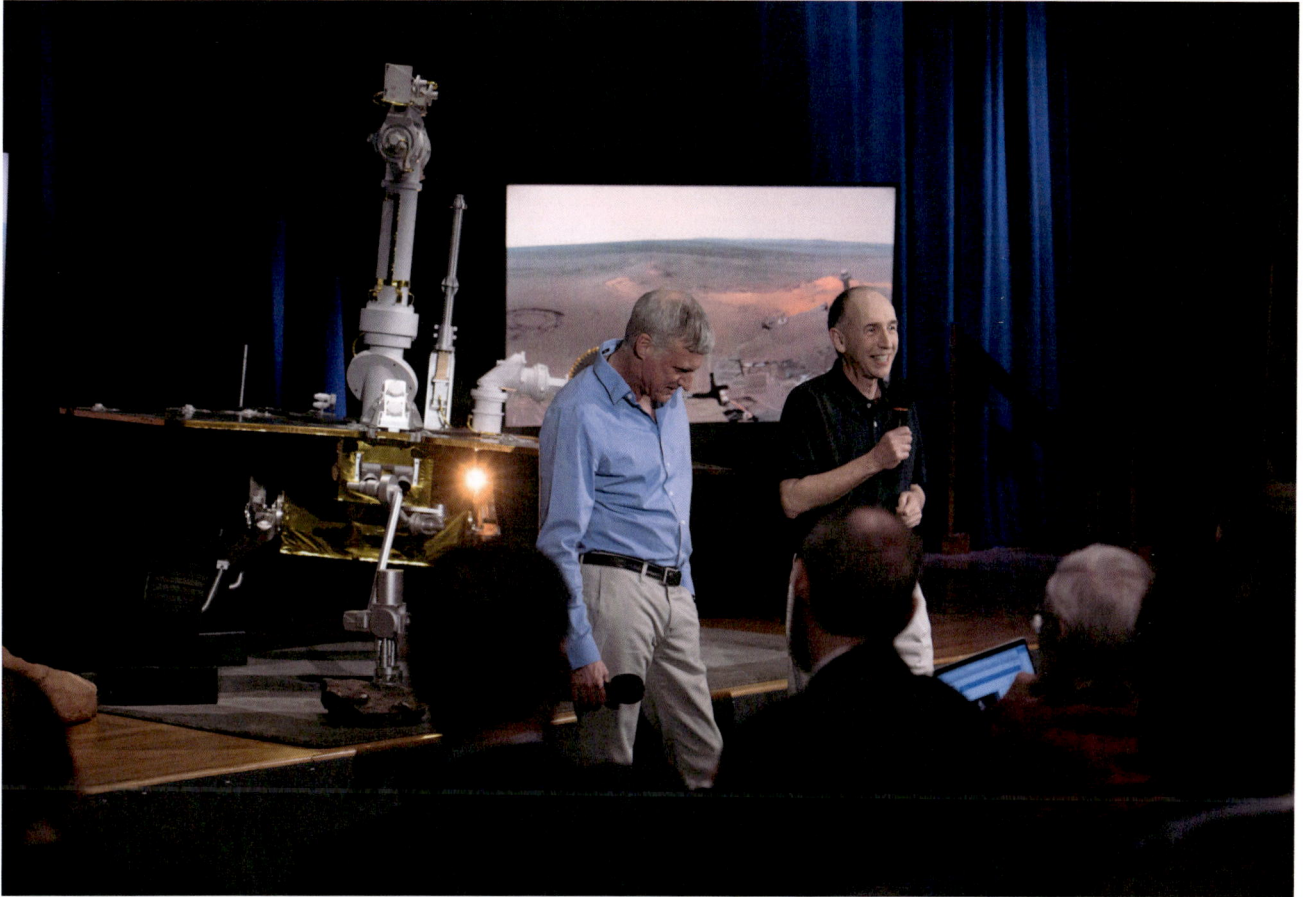

On February 13, 2019, Steve Squyres (left), MER principal investigator, and Matt Golombek (right), MER project scientist, stand in front of a model of the twin rovers and talk about Opportunity and Spirit.

Before being overcome by the storm, Opportunity sent a series of images to Earth.

Tabular rocks

Opportunity's entry point to Perseverance Valley

Endeavour Crater rir

Pitted rocks

Portion of solar panel

When combined, the panoramic picture shows a sweeping view of Endeavor Crater and Perseverance Valley.

A fitting name for Opportunity's final resting place.

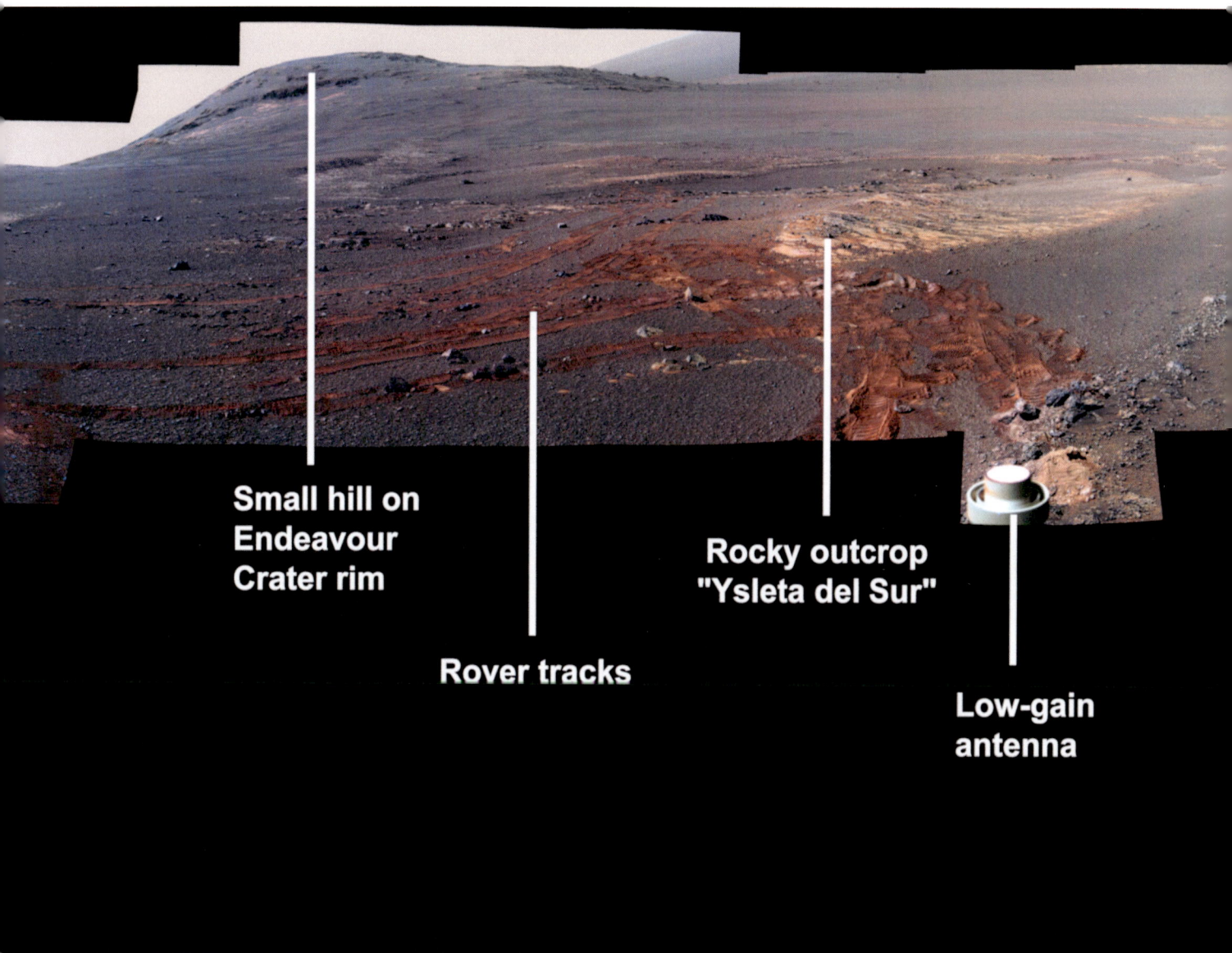

Small hill on Endeavour Crater rim

Rover tracks

Rocky outcrop "Ysleta del Sur"

Low-gain antenna

Perhaps on a future mission, it will be human hands that wipe away the Martian dust and find Opportunity.